B. P. Warwick

The Gas Engine

B. P. Warwick

The Gas Engine

ISBN/EAN: 9783741191855

Manufactured in Europe, USA, Canada, Australia, Japa

Cover: Foto ©Andreas Hilbeck / pixelio.de

Manufactured and distributed by brebook publishing software
(www.brebook.com)

B. P. Warwick

The Gas Engine

THE

Gas Engine

HOW TO

Make and Use it.

BY B. P. WARWICK.

ILLUSTRATED.

1897.
BUBIER PUBLISHING COMPANY,
LYNN, MASS.

CONTENTS.

The Gas Engine.

How to Make And Use It.

CHAPTER I.
History of the Gas Engine.

The gas engine is one of the wonders of the 19th century. Now, within three years of the 20th century, it is a novel machine, eagerly sought by many people. It is thought by persons who have not studied its principles that it is a steam-engine, using gas or gasoline as fuel for the purpose of making steam. This is erroneous. Gas and gasoline in specific proportion with air are explosive material.

The expansive force derived from explosion of these materials in the cylinder is the force that is substituted for the expansive force of steam. Hence, owing to the economy of this method as a means of deriving power, the steam engine and boiler are fast disappearing, and the Gas Engines taking their place for small power.

It is not generally known that the conception
of the gas engine for the production of motive
power antedates that of the steam engine.
Nevertheless such is the case. It was not until
the year 1744 that James Watt ran his first
successful steam engine at the Soho Works,
Birmingham, England, while Huyghens, Papin
and other scientists had produced power in
the seventeenth century by the explosion of
chemicals and the expansive force of heated
vapors. But the mechanical difficulties proved
too great, and so little was known in that age
of economical methods of producing gas as fuel
for the production of heat, that when the ap-
parently more simple method of using the expansive force of steam produced by the evapor-
ation of water by heat produced by the combustion of coal or wood was discovered, and Watt
constructed his engine for utilizing steam, all
further efforts in the direction of producing
power by other heated vapors were suspended.
For a century and a half the intelligence and
genius of the world have been expended in
improving the steam engine, until it is now
conceded that it is as perfect as human effort
can make it. No further economy in producing power by this method can be accomplished.

After careful and intelligent tests by experts
with the best instruments made at the present
day, it is generally admitted that what is now
deemed the perfect steam engine does not convert more than 10 per cent. of the heat efficiency
into indicated work, and that ordinary engines
and boilers do not realize over 4 per cent.

From 85 to 96 per cent of energy is lost in
the wasteful methods of producing steam
through a boiler, and the condensation and friction in conveying it to the piston of the engine
where the energy is expended in work. Com-
pare this with the effective energy produced by
the expansive force of heated vapor produced
by the combustion of gas in the cylinder of the
engine, without any intervening throttling by
friction, cooling and condensation. We have
immediately gained from 10 to 20 per cent. of
effectual energy from the heat units produced.

A century elapsed after the experiments of
Huyghens in 1680, and Papin in 1690, before
any further improvement or experiments are
recorded in this direction.

Soon after steam had been successfully used,
attention was again directed to the gas engine,
and we find that Robert Street obtained a
patent in 1794, and Lebon in 1801. In 1823

Samuel Brown constructed a gas engine that ran a boat successfully on the Thames and a road carriage on the streets of London.

W. L. Wright in 1835, and Barnett in 1838, made several important improvements. Then come Ador, Johnson, Robinson, Reynolds, Brown, Bolton, Webb, Newton, Edington, Barsanti, and Matteucci in the order named, who lend their aid in attempts to improve the machine. But it is not until 1860 that the mechanical difficulties are surmounted and a commercial gas engine constructed. Lenoir then brought the gas engine out of the experimental stage into public use, and the Reading Iron Works Co. of the U. S. built 100 of them, and several of these engines are in use at the present time. Next in line comes Hugon in 1865, and Otto and L. Langen in 1867, followed by McGreggor, Bulkeley, Clerk, Crossley, and Bisschop and others. Many years of constant application, labor, study, and experiment, with the usual discouragements and trials, have finally culminated in bringing out an engine that equals the finest steam engine of today. It is only a matter of time when the prejudice that, as usual, exists against any innovation, the ungrounded fear of explosions and other difficulties, will be overcome and the superiority of the gas engine over the wasteful steam engine and boiler will be established. The unsightly smoke stacks, belching forth smoke and soot, will be relegated to the scrap heap, and the iron will be used to construct more useful machines. The atmosphere of our manufacturing cities will be as clear as that of the country. The cost of producing power will be so reduced that the beggar may ride, and in the next decade the steam engine will occupy the same relative position to the gas engine that the flint and steel now do to the lucifer match, the tallow dip to the electric light, the stage coach to the bicycle, motor cycle, or the modern electric street cars, and civilization will record another grand stride toward the millennium.

Gas Engines may be divided into two types, those working without compression and those that have compression. In the former, the gas and air are drawn into the cylinder at atmospheric pressure and exploded or expanded, thus producing the power. In the latter, the mixture is first drawn into the cylinder, then compressed and later exploded. As this produces a better or more powerful explosion, the economy of gas by this method is apparent.

We will first describe a typical engine of the non-compression type, the Bisschop.

CHAPTER II.

The Bisschop Non-Compressing Gas Engine.

The Bisschop Engine was invented by Alexis De Bisschop in 1870 and was used mainly for small power, and although it is still made it can scarcely be called a modern gas engine. It appeared about four years after Dr. Otto and Professor Langen invented their non-compression atmospheric engine, and was intended especially to avoid the noise and recoil of the free piston, rack and clutch gear and the other defects of that motor. It belongs to what is called a mixed type. The charge of gas and air is admitted at atmospheric pressure, and the force of the explosion drives up the piston, but it is attached in a special way to the crank, and does not run free.

The pressure of the atmosphere, and the energy stored up in the fly-wheel then drive down the piston into the vacuum formed by the cooling of the gases. The action of the walls

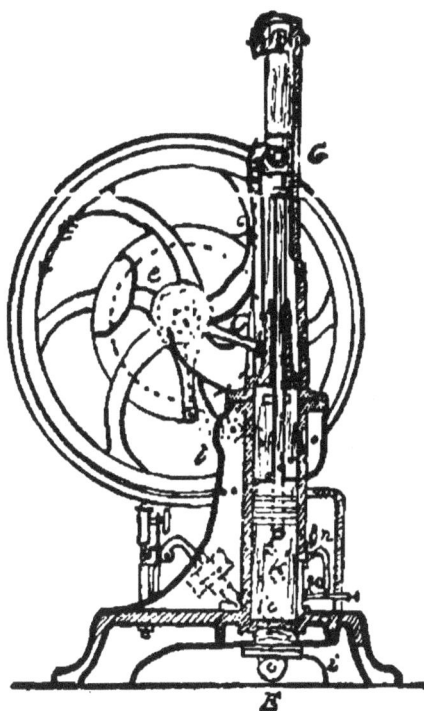

Fig 1.

Bisschop Engine
Sectional
          Elevation

FIGURE 1.

is here partly turned to good account, reducing
the temperature of the exhaust gases and helping
to form the vacuum.

In a certain sense the Bisschop, like other
atmospheric engines, may be called double-

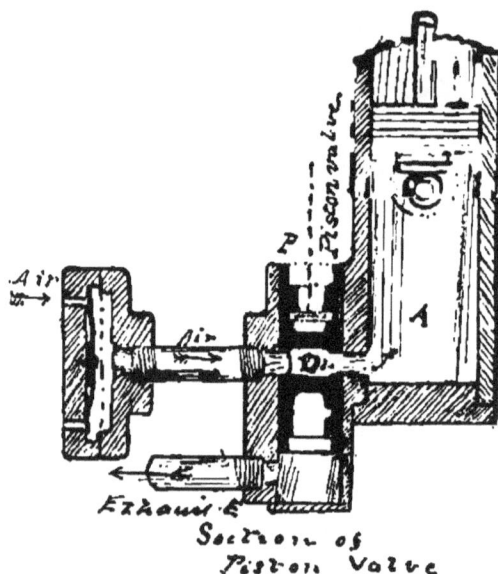

Section of
Piston Valve

FIGURE 2.

acting. The force of the explosion being used
on one side of the piston, and the pressure of
the atmosphere on the other; with the exception
of a few small French motors, it is probably the
only non-compressing engine still in the market.
It is said that about 2,000 of these engines were

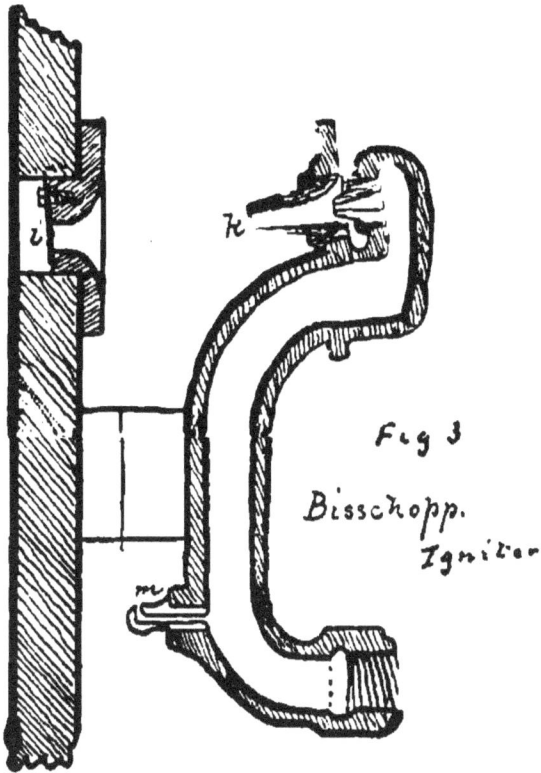

FIGURE 3.

sold in England during the first two years of
their being on the market.

Like all non-compressing engines the Bisschop is not very economical, and this may be
the reason why it is no longer in favor in places
where the high price of gas makes economy of
gas consumption a consideration.

Many cases occur, however, where simplicity and ease in starting and in handling are
necessary, and here the Bisschop, which is a
most convenient and simple little motor, has
been found of use for small powers.   For the
convenience of such of our readers as desire a
simple engine of this type we give a description
and drawings of the Bisschop.

The engine has a vertical cylinder, closed at
both ends, and the piston rod works in an upright hollow column.   Above is a cross-head
from which the connecting rod, working directly
through the crank on the motor shaft, hangs
parallel to the piston rod during the up stroke.
All these parts are close to the high column
carrying the piston and rod, and this causes a
certain amount of vibration, but the impulse
from the piston to the crank is direct.

Explosion occurs immediately after the piston
has passed over the lower dead point.   The

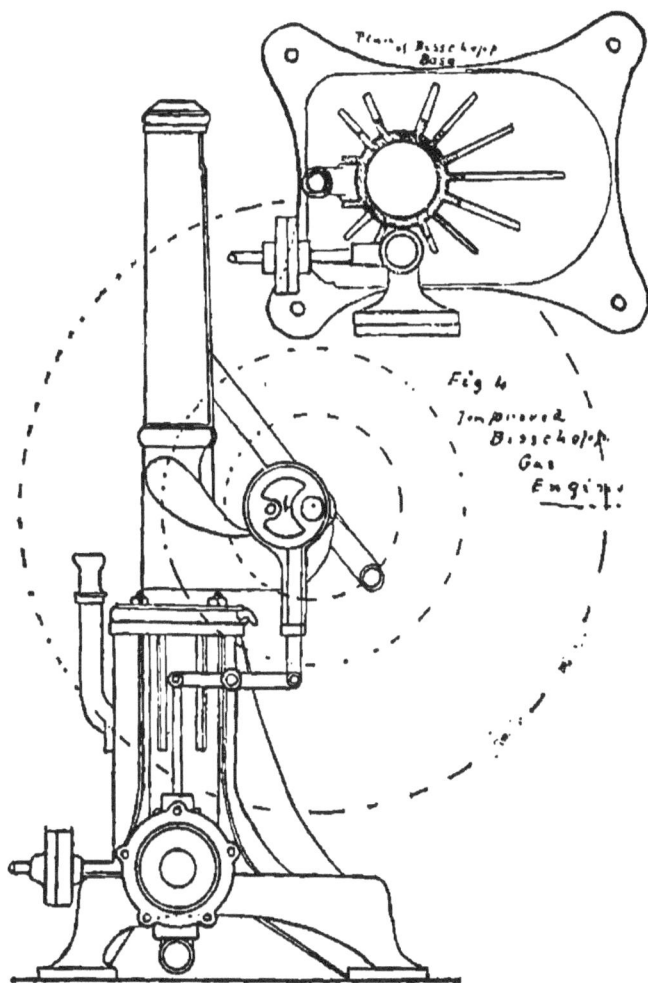

Plan of Bisschopp Bass

Fig 4
Improved
Bisschopp
Gas
Engine

FIGURE 4.

shock forces up the piston rapidly, the crank is
carried around through more than half a revolution, and the connecting rod brought parallel
with the piston rod inside the column.   Thus
expansion is exceedingly rapid and proportion-
ally greater than admission.   The distribution
of the gas and air, and the discharge of the exhaust gases, are affected by a trunk piston valve
driven from an eccentric on the crank shaft.
Gas and air are first admitted through valves
covered with thin rubber discs.   The air valve
is perforated with eighteen and the gas valve
with three holes admitting the charge in the
proportion of six parts of air to one of gas.

The piston valve is then driven down and
brought into line with the distributing chamber,
and the corresponding admission port of the
cylinder.

The engine has no water-jacket, the cylinder
being provided externally with ribs to cool the
metal.   Strange to say, it not only works without oiling, but the manufacturers expressly stip-
ulate that neither the piston nor the other parts
shall be lubricated.   A few drops of oil are
applied occasionally to the cross-head and the
motor crank only.   Ignition is obtained by an
external flame.   The piston valve admits, dis-

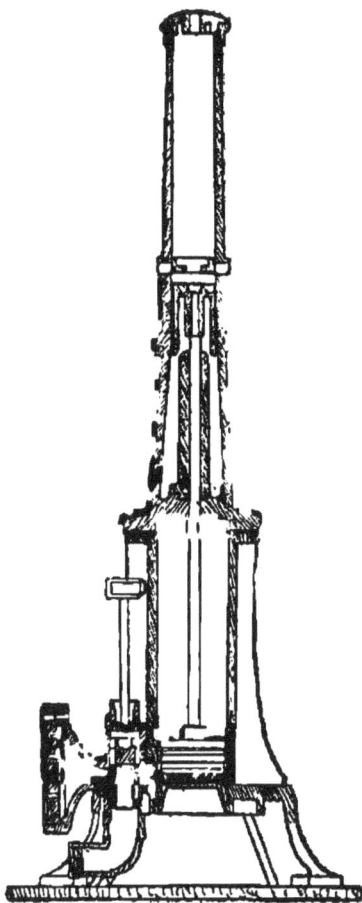

FIGURE 5.

tributes and expels the charge through the lever
1, (Fig. 1.) The exhaust is seen at E; K is
the small opening about half way up the cylinder, covered by a flat valve; an external flame
burns behind it at N, and at O is a second auxiliary flame, to rekindle the other when blown
out. Fig. 2 shows the air valve with the holes
for regulating the supply and the action of the
piston valve P; the gas enters at I (Fig. 1).

Method of Working.—Beginning with the
piston in its lowest position, when the exhaust
has just been cut off, the pressure in the cylinder being below the atmosphere gas and air
enter and mix in the distributing chamber.
The eccentric drives down the auxiliary piston
and brings its openings, M, opposite the mixing
chamber and the port F into the cylinder.
The charge enters, while the energy stored up
in the fly-wheel carries the piston past the
lower dead point. The opening at I is passed
and the flat valve hanging loose before it is
lifted by the vacuum, the flame is drawn in and
the charge fired, explosion follows, and the
pressure closes instantly the admission and
igniting valves, until the piston valve, raised by
the eccentric, has shut off the distributing
chamber, the piston flies up with great velocity

and more energy is generated than can be utilized in the up stroke. The reserve force
carries the fly-wheel through the remainder of
its revolution, and drives the piston down.
The exhaust valve is next open, and during the
greater part of the down stroke the gases of

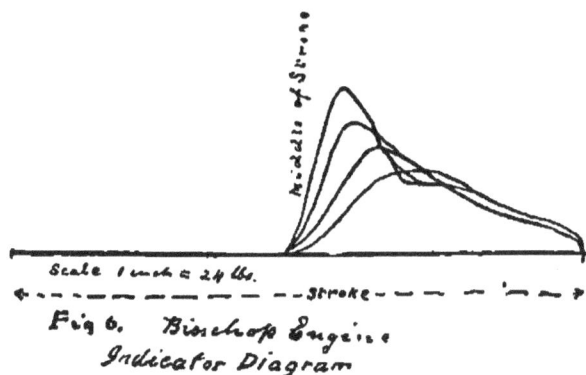

FIGURE 6.

combustion are driven out through the port uncovered by the piston valve which is now in its
highest position. When the pressure in the
cylinder is below atmosphere, and a vacuum
has been formed, the suction lifts the rubber
discs covering the gas and air admission valves,
the charge enters and the cycle is repeated.
The exhaust down stroke is a trifle slower than
the up expansion stroke.

This engine needs no governor, the regula-

tions of the speed being effected by (2) rubber
bags. The larger one acts as a reservoir and
the gas passes from it into the smaller bag,
which is so constructed that it receives and
passes on to the cylinder exactly as much gas at
a time as is required to keep the engine at any
given speed.

Test.—Several test experiments have been
carried out on this engine; all show a relatively
large consumption of gas. This engine should
not, however, be judged only by its expendi-
ture of gas; neither water nor oil are required
for the cylinder, and the motor is often used to
replace manual labor. Its advantages disap-
pear when the engine is made for larger powers, although the consumption of gas is propor-
tionately diminished. In England, where it is
most employed, it is seldom constructed for
more than one horse power.

Fig. 6 shows a diagram or card taken from a
one-horse power Bisschop Gas Engine.

CHAPTER III.

The Day Gas Engine.

Messrs. Day & Co., of Bath, England, have recently put upon the market an extremely simple and ingenious engine of the compression type, which, having been extensively copied with more or less modification in this country, is worthy of a short notice.

In this vertical motor several modifications from the general run of gas engines have been introduced, and although many of them have appeared in other engines, they are here utilized in a new and original way. With one cylinder only, an explosion is obtained at every revolution.

The cylinder and piston are at the top, and the latter works downward upon the crank through a connecting rod. Instead of a pump a reservoir is formed by enclosing the crank in an airtight chamber, and through a check channel or passage at the side the mixture is forced

from it into the upper part of the cylinder.
With the exception of this reservoir and charg-
ing passage, the mechanism of the Day Engine
is very simple.   There is no counter-shaft or
eccentric.   The action of the piston   itself

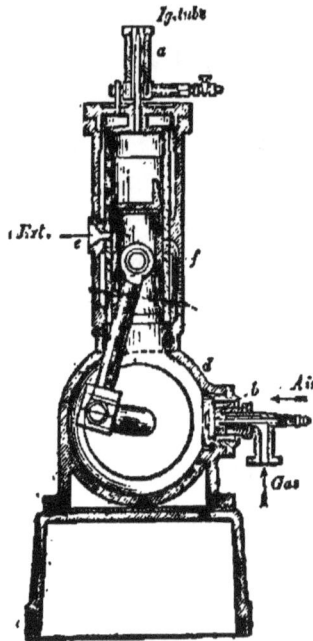

FIGURE 7.

causes the admission, explosion and discharge of
the gases.   There is only one valve, through
which the gas and air are automatically admitted in proper proportion by the suction of
the up stroke of the piston.   The exhaust gases

are expelled at the same time.   Ignition is by a
hot tube without a timing valve, placed at the
top of the cylinder.   As there is an. explosion
every revolution there is no danger of prema-
ture ignition, the gases being driven into the
hot tube at every up stroke by the compressing
action of the piston.

Fig. 7 gives a sectional elevation of a Day
Gas Engine.   A is the hot tube for ignition, B
the automatic valve for the admission of gas and
air, D is the chamber enclosing the crank, into
which the charge from B is first drawn.   At E
is the exhaust, which is merely an opening half
way down the cylinder, uncovered by the
piston.   F is the channel connecting the crank
chamber with the working parts of the cylinder.
All the four operations of the Beau de Rochas
Cycle—admission, compression, explosion plus
expansion, and exhaust—are performed in one
up and one down stroke of the piston, the down
being the motor stroke.

The action of the engine is as follows: the
crank being at the lower dead point and the
trunk piston at the bottom of the cylinder, its
edges just clear the port opening from the
channel F in the side into the upper end of
cylinder.   Through this channel during the

latter part of the down stroke, the fresh charge,
forced out by the piston, has been passing from
the reservoir D.   The up stroke commences,
and the port above F is immediately closed; the
upper face of the piston compresses the gas and
air above it, and drives them up into the ignition tube A.   Meanwhile, the reservoir having
been partly emptied of its contents through the
side channel, a partial vacuum is formed below

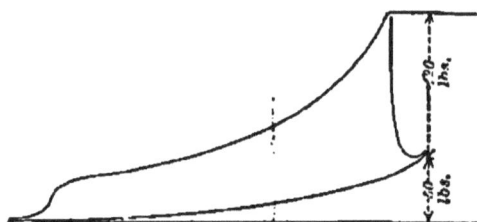

Fig. 8 :—Day Engine—Indicator
Diagram.

FIGURE 8.

the piston; the automatic valve B is lifted, and
a fresh charge enters and fills the reservoir D.
The piston having reached the end of the up
stroke the charge is fired, and the expansion
drives it down; the exhaust port is uncovered
and the gases discharged.   When the piston
is through half its stroke, it begins to force the
fresh charge in the reservoir below it through
the side channel into the upper part of the

cylinder before the exhaust port is covered, the incoming charge, already slightly compressed, helps to drive out the products of combustion. The return stroke compresses the mixture, and the cycle recommences. The simplicity of the Day Gas Engine makes it easy to reverse its direction of rotation.

Fig. 8 gives an indicator diagram or card of a $1\frac{1}{2}$ h. p. nominal engine, indicating 3.3 h. p. The diameter of the cylinder is $4\frac{1}{2}$ inches, stroke $7\frac{1}{2}$ inches, and it runs at 180 revolutions per minute.

CHAPTER IV.

The Sintz Gas Engine.

This engine, which is built by the Sintz Gas
Engine Co., of Grand Rapids, Michigan, is an
improvement on the Day Gas Engine of England, and is one of the few gas engines that is
suitable for Electric Light Work. Owing to the
excellent governor and also to its two cylinders,
giving it an explosion or impulse at every half
revolution, it is especially adapted to this class
of work. These engines are fitted with either
a hot tube or an electrical igniter, either of
which gives excellent results. Fig. 9 shows a
cut of this engine.

The Monitor Gas Engine and the Wolverine
Gas Engine, both of which were constructed in
Grand Rapids, Mich., by companies started by
Mr. Clark Sintz, are both of the same type
with perhaps a few modifications, as the Sintz
and Day engines, and give excellent results, and
are remarkably economical as to the amount
of gasoline they use.

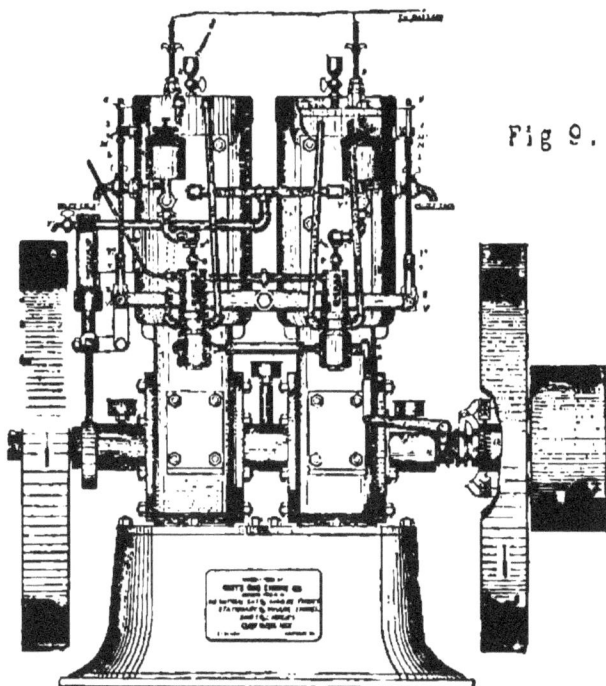

Fig 9.

FIGURE 9.

CHAPTER V.

The Olin Gas Engine.

The Olin Gas Engine, manufactured by the
Olin Gas Engine Co., of Buffalo, N. Y., is of
the four cycle type and lays claim to various
novel features. Chief among these is a simplicity of construction heretofore unknown in
gas engine manufacture; and a gas consumption of not more than fifteen feet per brake
h. p. per hour. The charge is introduced
through a piston valve-like tube, with a grind-
ing face at end, to which gas and air suction
valve is seated. The tube itself, with a self-
adjusting collar and a ground seat in the valve
chamber, serves as an exhaust valve, and is
actuated by a cam on the main governor gear.
The governor consists of two weights accurately
balanced and attached to a finger, this latter
engages, when the speed of the engine drops,
with a dog controlling the travel of the exhaust
valve, and attached to a rock lever. During
accelerated speed, the finger is depressed and
the dog rides on the governor disk, holding the

FIGURE 10.—THE OLIN GAS ENGINE.

exhaust valve open and preventing the intro-
duction of a new charge.   The construction of
the governor, with its constantly moving weights,
insures close regulation.   The easy accessibility
of every part of the completed engine is also a
notable feature.

CHAPTER VI.

How to Make The Warwick Gas Engine.

The Warwick Gas Engine has been designed to meet the requirements of all who require a light power which is always ready and can be used as wanted. The cost and method of construction have been reduced and simplified to enable the average amateur to construct it.

The Warwick Gas Engine belongs to the class of engines exploding at constant volume with previous compression. The working cycle is that known as the Beau de Rochas Cycle, and is divided into four parts in which the engine makes two revolutions. During the first complete revolution of the engine the cylinder acts as an air pump. As the piston moves forward gas and air in the proportion of six parts of air to one of gas are admitted through the automatic inlet valve. When the piston has reached the forward or top end of cylinder the inlet valve closes, and as the piston returns to the back end of cylinder the charge of air and gas is compressed to about one-third its original volume.

At the beginning of the second revolution the compressed charge of gas and air is ignited and exploded, either by an electric spark, a hot tube, or a flame, all of which methods are shown, it being left to the discretion of the builder as to which method is adopted.

The explosion forces the piston forward until it reaches the front or top end of the cylinder, at which time the exhaust valve opens, and during the return stroke the burnt gases are discharged through the exhaust pipe.

As will be seen, the engine receives one impulse in every two revolutions, and during the first revolution no power is developed, but on the contrary it is absorbed or destroyed, by the compression.

This, viewed from the steam engine point of view, seems on a par with the man who attempted to lift himself by his boot straps, but in spite of this apparent waste of energy, the gas engine is about three times more efficient than the steam engine.

As the working drawings are not only made to scale, but have also dimensions plainly

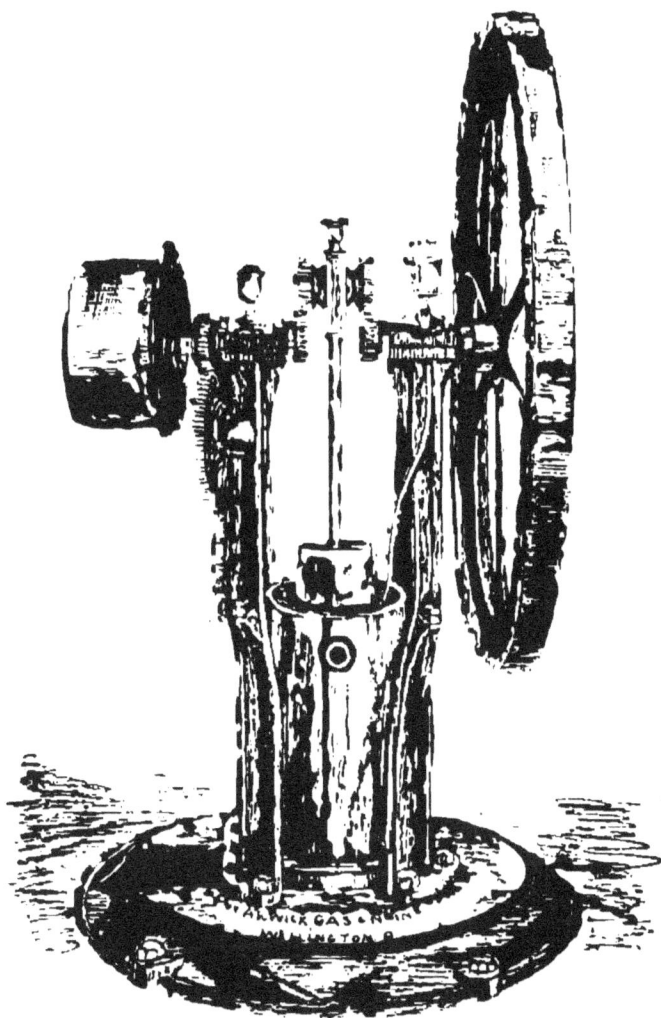

FIGURE II.—THE WARWICK GAS ENGINE.

marked in figures, there is little necessity for an
extended explanation.

A few special pointers to be used in connection with the working drawings.

The Cylinder.—The cylinder is surrounded
with a water jacket $1\frac{1}{4}$ inch thick; the ribs
around the cylinder under the jacket form
channels through which the water is circulated,
cooling the cylinder and being finally discharged at the top.

After boring the jacket, place the cylinder on
a mandrel, (having first bored the cylinder out
four and a half inches) and turned down the
flanges until they almost fit into the jacket; heat
the jacket to a bright cherry heat and it will
expand enough to pass over the flanges, and on
cooling will shrink, making a tight joint.

Before shrinking on the jacket, carefully note
the exact location of the bosses cast on the sides
of the cylinder, these are to be drilled and
tapped and $1\frac{1}{2}$ in. 10-24 screws put in to firmly
hold the jacket in position; again mount the
cylinder on mandrel and turn off the outside of
jacket so that it will fit inside of ring in the

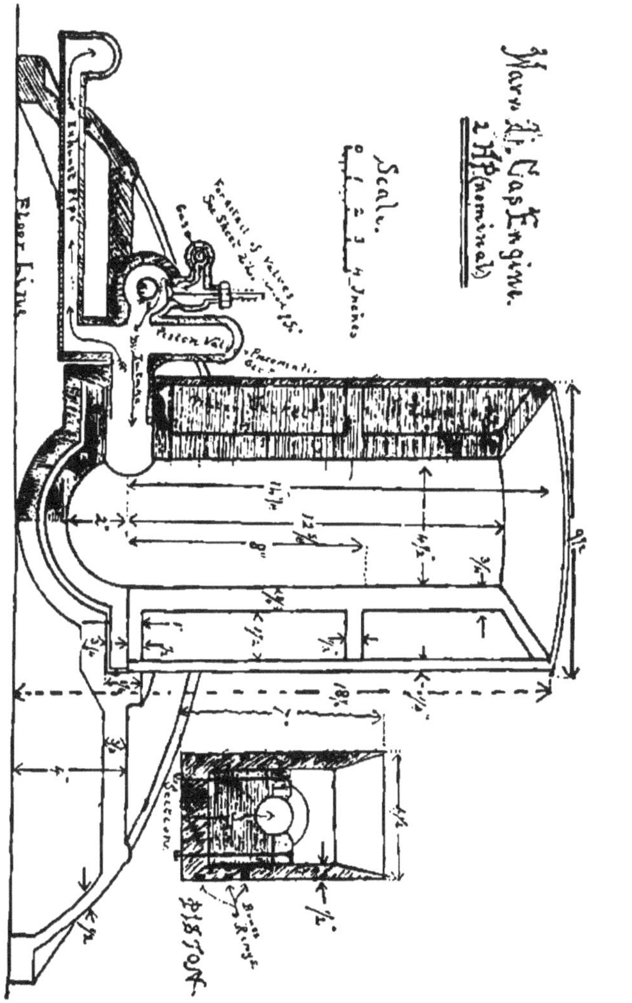

FIGURE 12.

base.   Now face up cylinder head both sides,
and having made a good joint hold it in place
with four half-inch, counter-sunk-head screws
one inch long.

The cylinder is now ready to be mounted on

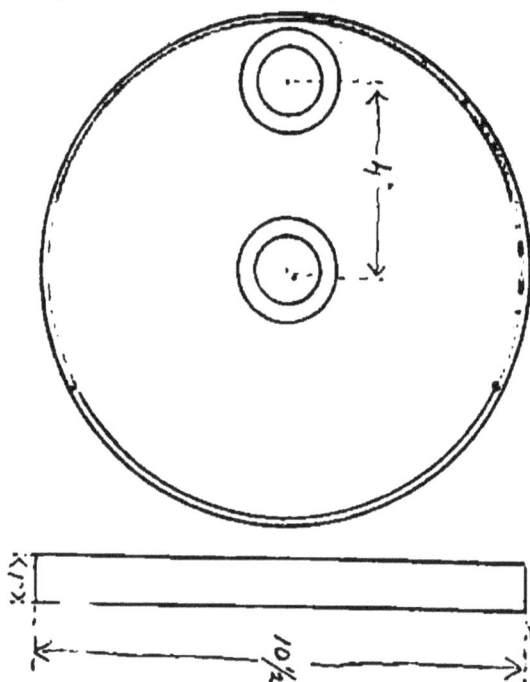

FIGURE 13.—CRANK DISK.

the base, where it should be secured with at
least six half-inch cap screws.   A thin layer of
asbestos, or even white lead, may with advan-
tage be placed between the cylinder head and

the base before screwing down. The exhaust
and intake ports should be tapped out, and the
combination valve screwed on.

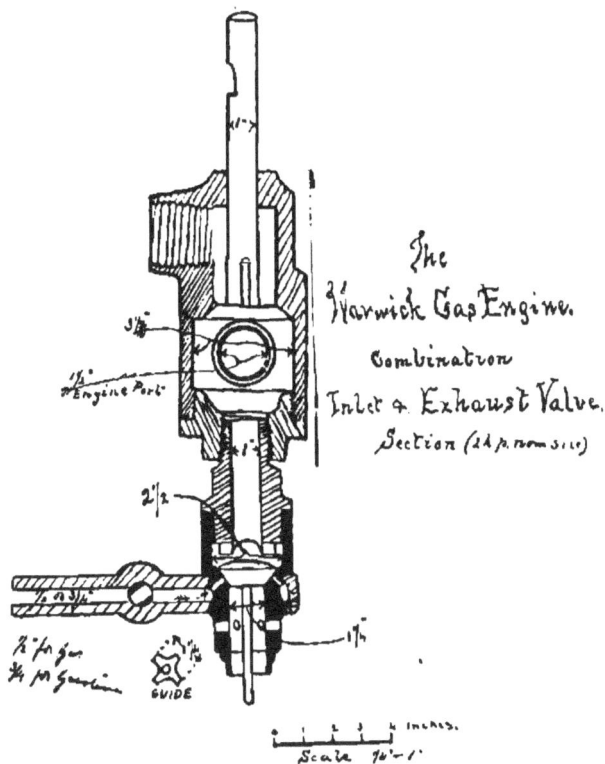

FIGURE 14.

Valves.—The builder here has the choice
given to him of two valves, one with a pneu-
matic and the other with a centrifugal governor, either of which gives excellent results, but

require to be carefully made. Both exhaust,
gas, and air inlet valves are made conical, and
should be ground in after the yare turned to size,
as a good deal depends on having tight valves if
you want the engine to work successfully.
The stems of both valves are of steel, and that
of the exhaust valve should be hardened after
the notch for the governor spindle has been cut.

Governor.—The action of the governor will
be apparent from the working drawings. The
lower end of the governing spindle should be
hardened where it comes in contact with the
notch cut in the exhaust valve-stem.

Water Jacket.—If you can make connection
from a water faucet to the cooling jacket of the
engine do so, or if this cannot be done a tank
can be used for the water to circulate in. If,
however, you intend to use the engine in a boat
it will be best to make the pump shown in the
drawings, and force the water in at the bottom
of the jacket, as that tends to keep the cylinder
cool and the water jacket full.

Battery and Spark Coil.—Where it is de-
cided to use an electric ignitor, it is advisable to
use a battery of three storage cells, (for prefer-
ence the P. W. Dry Storage Battery) as it will
not slop and has extremely high E. M. F.,

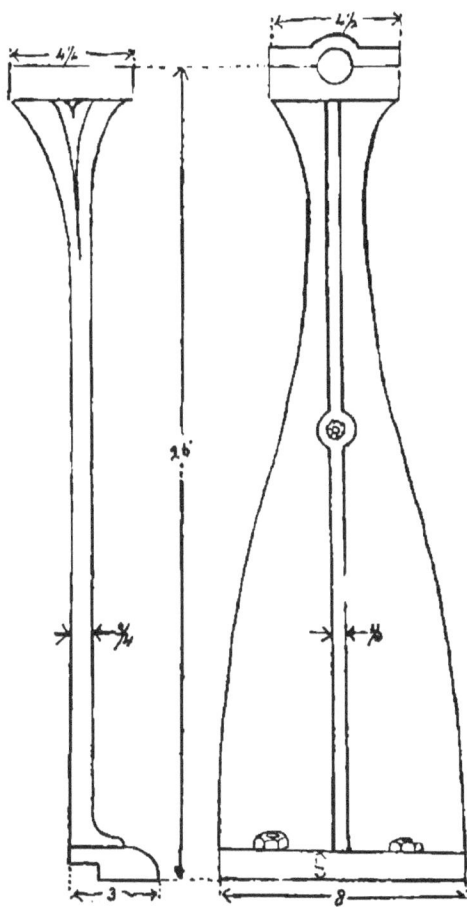

FIGURE 15.—PEDESTALS.

which tends to a fat or hot spark which is very desirable.   When it is not convenient, however, to obtain these cells, almost any good, large-size carbon battery will do if you have a good spark coil.   If a sal-ammoniac cell is used it is advisable to connect them, two in parallel or multiple, and say four or five in series.

The insulating bushing that passes through the cylinder head can be made of lava or porcelain as most convenient, and can be made to order by any pottery, or the lava can be obtained from the D. M. Stewart Mfg. Co., Chattanooga, Tenn.

Carburetor.—Where it is desirable to operate the engine with gasoline it will be necessary to have a carburetor, a sketch of which is shown and which needs no further explanation.

Ignition Tubes.—These may be obtained from the Fairbanks-Morse Co., from their factory at Beloit, Wis., or at one of their agencies.

CHAPTER VII.

The Wadsworth-Warwick Marine Gasoline Engine.

For the benefit of those of our readers who wish to use a gasoline engine for propelling a boat, drawings are given for a marine engine and to economize space a table has been prepared giving the exact dimensions of one-half, one-fourth and one h.p. engines. This table is published by permission of the Wadsworth Machine Works of Wellington, Ohio., who have given permission to use their drawings. Thinking that this hand-book will likely come into the hands of some practical machinists and engine builders who may desire to build a large size gasoline engine, scale drawings are given of a 25 h. p. (actual) engine designed by the writer several years ago in Europe and especially adapted, owing to its sensitive governor, to electric light work. Several were made to order for this purpose and have given excel-

lent results. As the drawings explain them-
selves to any mechanic no more need be said.
This engine is called the Simplex.

GENERAL DIMENSIONS OF STANDARD GASOLINE ENGINES.

THE WADSWORTH MACHINE WORKS.

TYPE M.

H. P.	A	B	C	D	E	F	G	H	I	J
$\frac{1}{4}$ · ·	$\frac{1}{4}$	$\frac{1}{4}$	$\frac{3}{16}$	$\frac{3}{16}$	$1\frac{3}{4}$	$1\frac{1}{4}$	5	$\frac{5}{8}$	$\frac{3}{16}$	$\frac{5}{8}$
$\frac{1}{2}$ · ·	$\frac{1}{4}$	$\frac{1}{4}$	$\frac{1}{4}$	$\frac{1}{4}$	$2\frac{1}{2}$	$1\frac{1}{2}$	9	$\frac{3}{4}$	$\frac{5}{16}$ ·	1
1 · ·	$\frac{3}{8}$	$\frac{3}{8}$	$\frac{3}{8}$	$\frac{3}{8}$	$3\frac{1}{2}$	2	13	1	$\frac{3}{8}$	$1\frac{1}{2}$
$2\frac{1}{2}$ · ·	$\frac{3}{4}$	1	$\frac{3}{4}$	$\frac{5}{8}$	5	$2\frac{1}{4}$	$21\frac{1}{4}$	2	$\frac{3}{4}$	$2\frac{1}{4}$

H. P.	K	L	M	N	O	P	Q	R	S	T
$\frac{1}{4}$ · ·	$1\frac{3}{4}$	$2\frac{1}{4}$	$\frac{1}{2}$	$\frac{3}{8}$	13	8		$1\frac{5}{16}$	$1\frac{5}{8}$	2
$\frac{1}{2}$ · ·	2	4	$\frac{3}{4}$	$\frac{1}{2}$	18	12	$10\frac{1}{2}$	$\frac{3}{8}$	$\frac{3}{8}$	3
1 · ·	3	$5\frac{1}{4}$	1	$\frac{3}{4}$	24	18	$15\frac{1}{2}$	$\frac{1}{2}$	$\frac{5}{8}$	4
$2\frac{1}{2}$ · ·	4	$8\frac{1}{2}$	1	$\frac{3}{4}$	36	24	21	$\frac{3}{4}$	$\frac{3}{4}$	5

The Warwick Simplex Gas Engine is shown
in Figs. 16, 17, 18 and 19. Fig. 16 being
the rear end elevation showing valves, etc.
These are shown again in Fig. 19. As the
principal feature of this engine is its extremely
simple and sensitive governor. We will give a
brief description as it can be adapted to any
engine.

Fig. 19 is a sectional plan and shows the

slide valve S having an admission port e lead-
ing to a mixing chamber M, to which the gas is
admitted from the gas inlet g, by the spring
poppet valve s, which is held closed by the

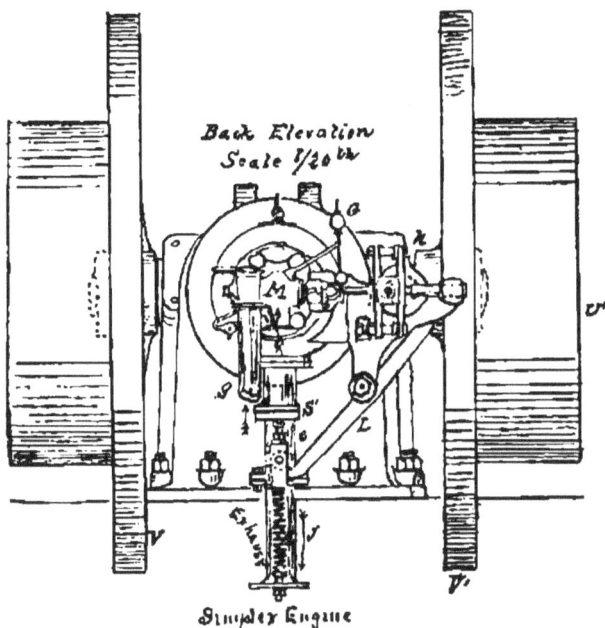

FIGURE 16.

spring H.   The slide valve S has a 1 to 4
motion transmitted to it by the small crank K
mounted on the valve rod R (Fig. 17).   The
admission and quantity of gas is controlled by
the small air pump (Fig. 19) c being the cylinder in which the plunger O works.   K is a

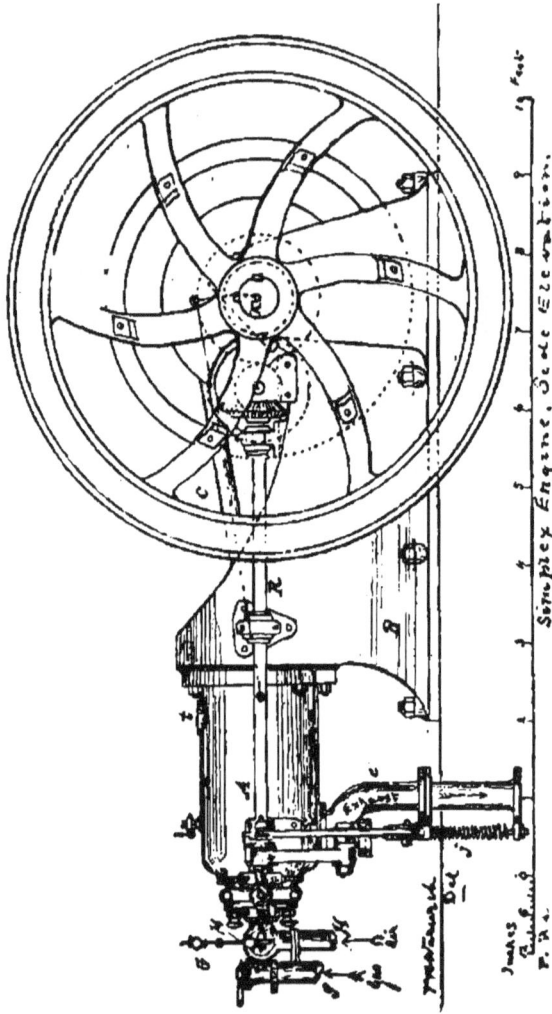

FIGURE 17.

Simplex Engine, Scale Elevation.

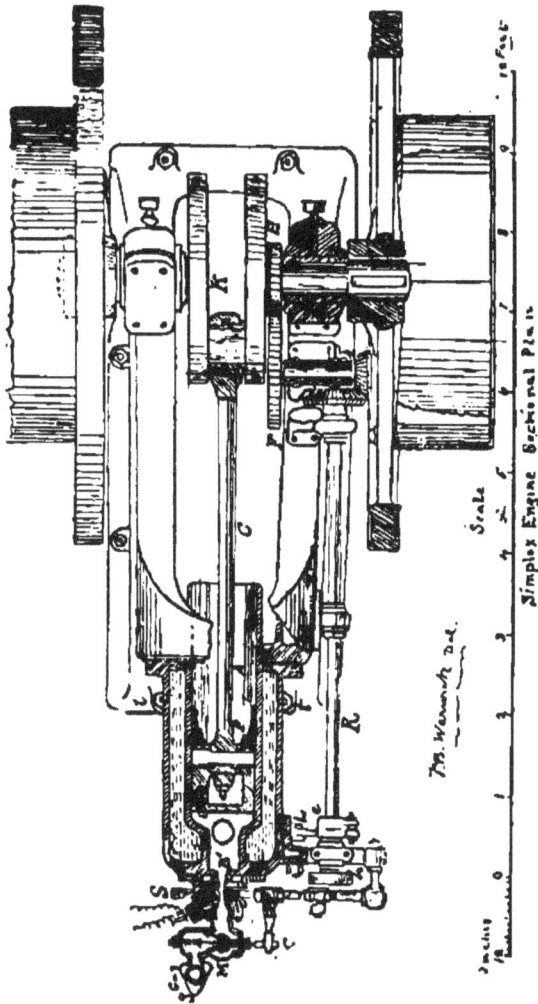

Simplex Engine Sectional Plan

FIGURE 18.

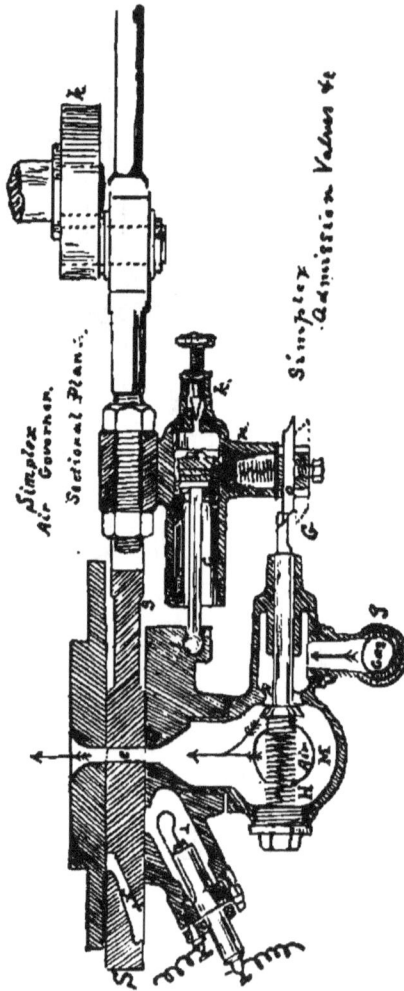

Simplex Air Governor. Sectional Plan.

Simplex Admission Valve &c.

FIGURE 19.

needle valve allowing the compressed air to
pass out of the air chamber. N shows the
plunger valve and spring carrying on its end
the trip finger o, which engages with the stem
G of the poppet valve S and forcing back the
spring H, allows the gas to enter the mixing
chamber M.

Should the speed of engine exceed what it is

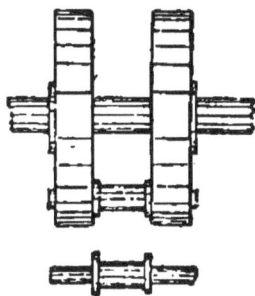

FIGURE 20.--CRANK PIN.

set for the air pump compresses air in the air
chamber more rapidly than it can escape by the
needle valve K, and the plunger N is forced
out and the trip finger o takes the place shown
by the dotted lines Fig. 19 and does *not* strike
the valve stem G and the poppet valve S is held
shut by spring H and an explosion missed.

The mixed and compressed gases are ignited
at the proper instant by being forced down the

port f, into the chamber r, where a continual stream of sparks is kept up by a small induction coil, or if preferred a platinum wire may be kept white hot by a battery or the regular hot tube may be used with good results. In Fig 17 A is the cylinder with its trunk piston at P. C the connecting rod and K the cranks. The method of building up the cranks by a crank pin being shown at Fig 20.

Wadsworth-Warwick
GAS ENGINE.

P.B.Warwick.
Del. — — ALL RIGHTS RESERVED. —

~1897~

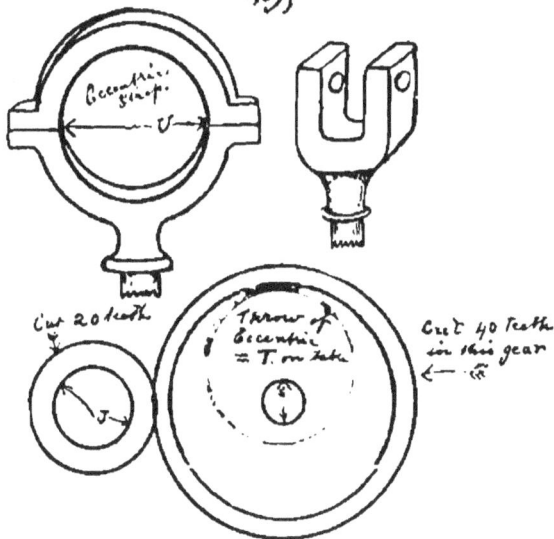

Eccentric strap.

U

Cut 20 teeth

J

Throw of Eccentric = T on tab.

Cut 40 teeth in this gear

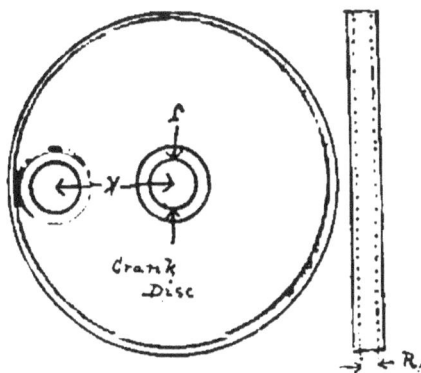

Y

Crank Disc

R.

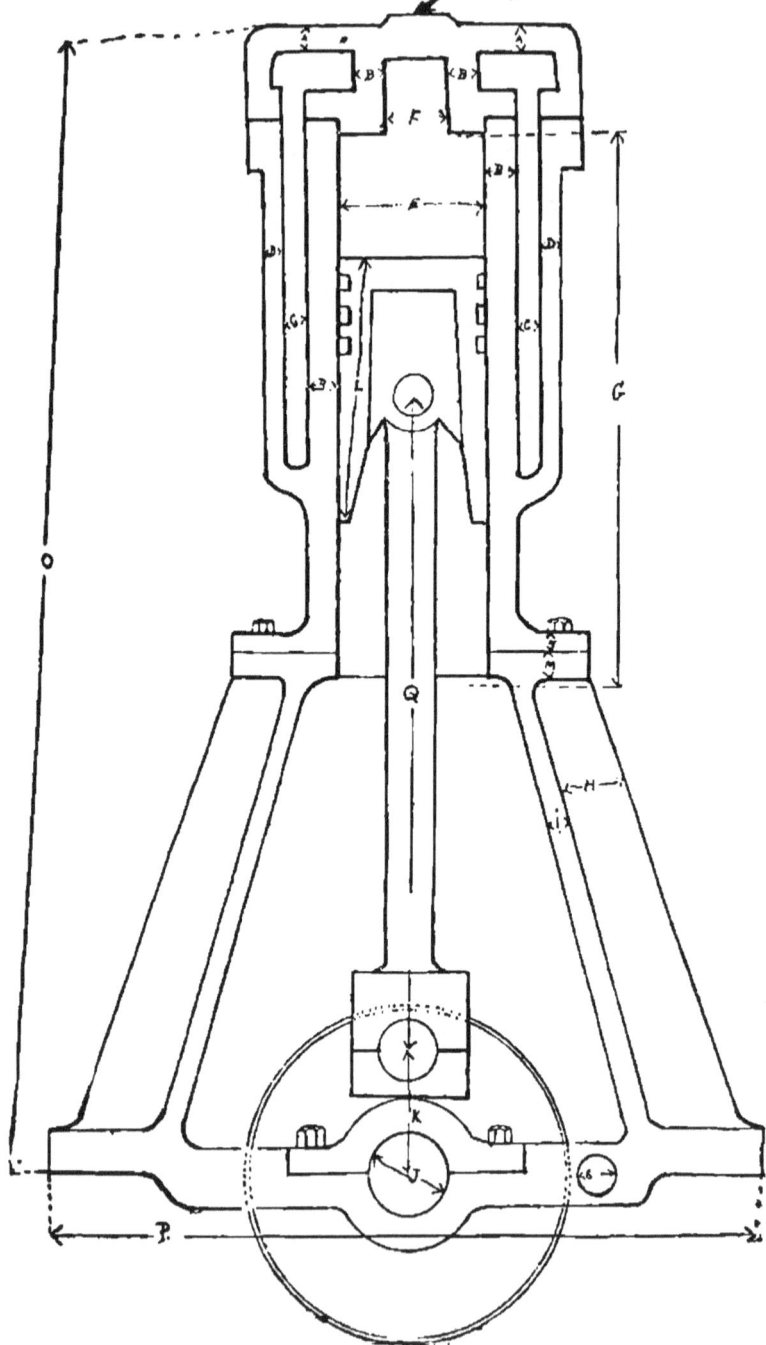

General Dimensions
Type M. Sheet A /47.

Igniter or Hot Tube releases in here

CHAPTER VIII.

The Fairbanks-Morse Gas and Gasoline Engine.

The Fairbanks-Morse gas and gasoline engines as shown in Fig. 21 are probably the best known of American Gas Engines, and are made in various sizes from 2 to 75 h. p. One of the principal features claimed by the makers is the simplicity of the mechanism, the number of the working parts being reduced to a minimum. Modern steam practice has been followed in placing the governor, which is located in the flywheel, thus dispensing with belts, gears, etc. The hot tube, or electric igniter, or both, is used as desired. Referring to the gasoline engine, this fluid is pumped continuously into a small brass container or reservoir on the engine, in which it is kept at a constant level. Air passing through a nozzle connected with this reservoir takes up sufficient oil to form an explosive

FIGURE 21.

mixture; this is regulated by the governor.
When an explosion is not required, the exhaust
valve is held open and no charge is admitted;
and the engine is relieved of the work of compression. There are only two valves on the
engine, both of which are of the poppet type
and well water-jacketed.   There is but one
cam which works the exhaust valve through a
straight rod. On all engines above 3 h. p. a self
starter is fitted which is very simple in construction and works extremely well.   The Fairbanks-Morse Co. have extensive works at
Beloit, Wis.

This engine was originally designed for
pumping in the west where irrigation is a
necessity, but met with such success that its
designor, Mr. Charter, who has been termed the
daddy of the American gasoline engine, modi-
fied it, and under its new name it is meeting
with great success, even for electric lighting,
which is the most severe test that can be applied to a gas or gasoline engine.   A plant is
now being designed by the writer where it is
intended to use several of these engines of large
size for incandescent lighting.   One of the
noticeable features of this engine is that even in
the large sizes not more than a pint of gasoline

is in the supply tank in the building at a time
and the reservoir containing the main supply is
in the earth below the engine and outside the
building, thus doing away with any objection
that the fire underwriters might have to the use
of gasoline in a building.   No carburetor is
used, the gasoline being sprayed directly into
the cylinder on the out stroke of the piston, the
exhaust being closed ; in other words the gasoline is inhaled into the cylinder and then compressed and afterwards exploded.

CHAPTER IX.

The American Otto Gas Engine.

The Otto Gas Engine, invented in 1867 by Dr. N. August Otto, and which was exhibited by him at the Paris Exhibition of the same year, and obtained a gold medal for general excellence, is manufactured in this country by The Otto Gas Engine Works of Philadelphia, Pa., who state that over 47,000 engines, representing over 250,000 h. p., have been sold and are still in use. The writer, while in the employ of Messrs. Crossley Bros., of London and Manchester, England, in 1883, set up one of the first upright or domestic Otto engines in Otto & Crossley's window on Cheapside, London, and connected it direct to a Siemens incandescent dynamo. It suffices to say that the engine and dynamo are still there, working every day lighting the offices of the concern. Messrs. Crossley Bros. have recently installed several large engines (one of 350 ind. h. p.) in Ireland for electric lighting, which is a guarantee that these engines are perfectly regular and govern very closely. A cut of this engine is shown in Figure 22.

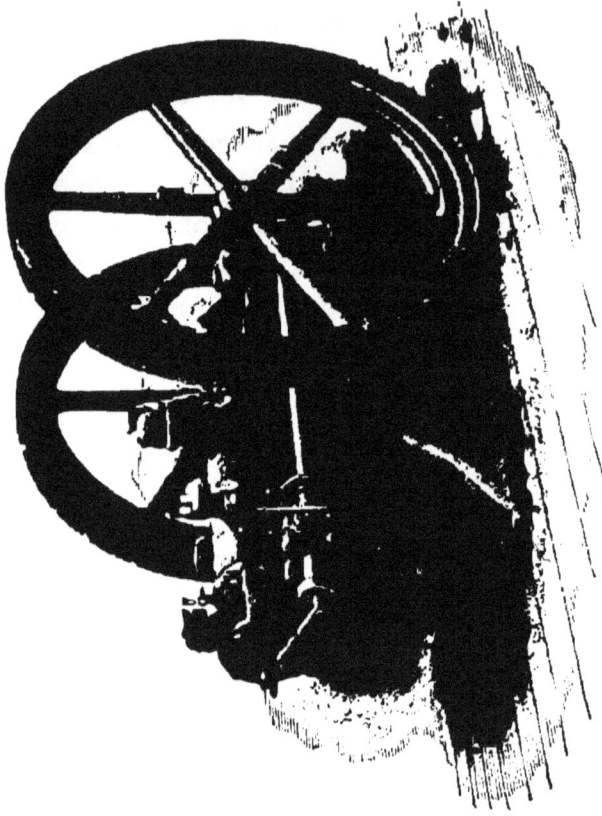

FIG. 22.—THE OTTO GAS ENGINE.

CHAPTER X.

How to Make a Carburetor for a Gasoline Engine.

In many places it is impossible to obtain gas to operate the engine, therefore we will consider how a carburetor can be made. While almost any method of saturating air with gasoline will do, even the crude method sometimes employed of forcing the air through the liquid giving fair results, still this is not economical, as, if the gasoline is heavy or common, only the more volatile oil will be vaporized and the balance must be thrown away. A cut is shown of a very simple carburetor by means of which even coal oil (kerosene) can be used. The apparatus consists of two parts, the heater and the carburetor, or saturator. The latter is shown at Fig. 23 in section. It is a simple cast-iron shell with a float working in an inside chamber.

1. Is the float.
2. Float tube cap.
3. Screen.
4. Plug.
5. Tube cap screw.
6. Sight glass.
7. Screw ring.
8. Drain cock.
9. Holes admitting gasoline to float.
10. Horschair.
11. Level of gasoline.
12. Gasoline needle valve.
13. Shield.
14. Hot air inlet from heater.
15. Gas or vapor outlet.
16. Gasoline inlet from tank.

The heater is simply a chamber into which the exhaust passes on its way out to the open air, and is made by screwing two caps onto a piece of S-inch pipe and then drilling and tapping two holes in the ends so as to allow a piece of 1-inch pipe with two elbows and a nipple on it to be placed in it to make a heating coil. This is open at one end, and the other end connects to the hot air inlet No. 14, shown in

NOTE.—The diameter of the Carburetor for a 2 or 3 h. p. engine should be 9 x 11 inches high inside measurement.

Fig. 23. The action is as follows: the piston
on the out stroke draws in a mixture of gasoline
gas and air. The gas is generated by the air
drawn through the heater pipe being heated by

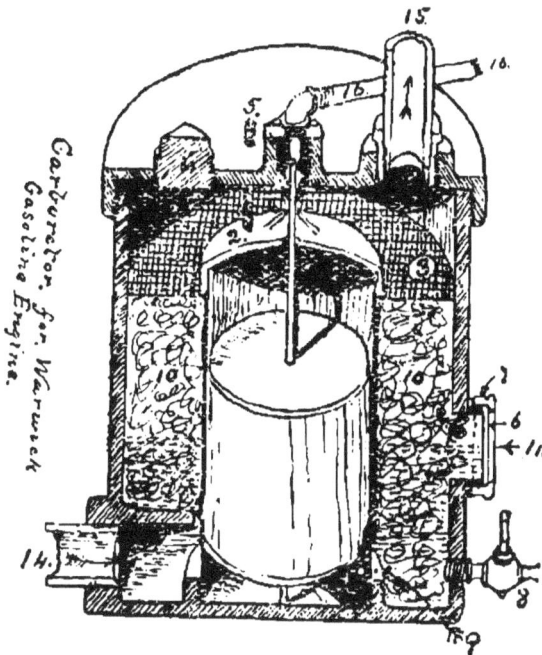

FIGURE 23.

the exhaust. This hot air, by the suction of the
piston, is caused to flow in at the hot air inlet,
and after bubbling through the gasoline in the
carburetor passes out of the vapor outlet, 15,
to the gas valve of the engine, is there
compressed and exploded, and then serves to
heat the air for the next charge.

CHAPTER XI.

How to Make a Simple Electric Igniter.

First procure a porcelain or clay wall insulating tube from any electric supply store. This should have a $\frac{3}{8}$-inch hole through the center and be about 3 inches long over all and have a $\frac{1}{2}$-inch head 1 inch in diameter. Now procure a piece of $\frac{7}{8}$-inch brass tube $1\frac{1}{2}$ inches long, and after cutting a thread on the outside of it place it over the outside of the wall tube filling up the space between with plaster of paris or asbestos cement. While this is setting, take a piece of brass rod $\frac{3}{4}$ inch in diameter and turn it down to a snug fit to the inside of tube. Make this three inches long and leave a collar on one end to prevent it going through the tube by the force of the explosion. Now turn up a collar $\frac{1}{2}$ inch by $\frac{5}{8}$ inch diameter and shrink it on to the end of the brass plug that projects through the insulating tube. Now drill a $\frac{1}{8}$-inch hole through the brass plug and this part is ready to

screw into the cylinder head where a suitable
hole must be drilled and tapped to receive it.
You will now require a short piece of drill rod
$\frac{1}{4}$-inch in diameter and 6 inches long, cut an
8-32-inch thread onto each end of this and fit
it to work nicely through the $\frac{1}{4}$-inch hole in the
brass plug. When this is done a piece of
round brass two inches long by one inch diameter should be screwed on to the inside end to
see how it fits and then removed and a $\frac{3}{8}$ or $\frac{7}{16}$
inch hole drilled in a little way, say one inch,
to receive a piece of carbon of the same size,
which may be secured there by a couple of 8-32
set screws. Now obtain a piece of fiber, or
better, vulcabeston, three inches long by $\frac{3}{4}$-inch
wide and $\frac{1}{2}$ inch thick, drill and tap an 8-32
hole through the center and one at each end to
receive the ends of the two springs which
should be made out of piano wire wound on a
mandrel. The igniter is now finished and
ready to be assembled and screwed into the
cylinder head and connected with a spark
coil and battery. The action is as follows :

The piston just before the end of the up or
compression stroke strikes the end of the carbon
rod and pressing up the steel rod stretches the
springs slightly by the momentum stored in the

fly wheels, the piston passes the dead point and starts on the down stroke, and then breaks contact with the carbon and piston head and a spark is drawn which ignites or explodes the compressed mixture, and an impulse being given to the piston the working cycle commences.

Nearly any good spark coil can be used, but a twelve inch coil such as is used for electric gas lighting is best.

NOTE.—Our hand dynamo, price $3.50, will be much better than batteries. You will need no spark coil with it.

BUBIER PUBLISHING CO., Lynn, Mass.

CHAPTER XII.

The Hornsby-Akroyd Oil Engine.

This engine has one peculiarity which distinguishes it from the other types of heat motors. This is the fact that it has neither a hot tube, an electric spark nor a slide-valve with flame to explode the charge of oil and air, and is therefore, perhaps the simplest oil engine now built. A peculiar feature is that no attempt is made to vaporise the oil, or convert it into spray until it is actually injected into the combustion chamber. Hence the density of the oil is a point of no importance and heavier petroleum may be used than in most other engines. The gravity of the oil used is usually about 0.850 and its flashing point 150 degrees F., but excellent results have been obtained with oil having a specific gravity of 0.864 and a flashing test or point of 225 degrees F., thus securing perfect safety in operation. The chief objections to gas and oil engines in general is the necessity of

providing artificial means of igniting the charge.
This objection is obviated in the Hornsby-
Akroyd type. The manner in which the explosion is obtained in this engine is as follows:
The cylinder is provided with an extension
communicating with it by a relatively narrow
neck. This extension is *not* water-jacketed and
forms a retort in which the oil is vaporized.
Nothing but oil in liquid form is injected into
the retort and only air is drawn into the cylinder. This engine is shown in Figure 24.

On the out stroke of the piston air is drawn
into the cylinder and oil is forced into the hot
retort. At the end of the stroke there is in the
retort oil vapor, which is not in itself explosive,
and in the cylinder pure air which is also non-
explosive, and there is not sufficient leakage
from one to the other to make either charge explosive. On the return stroke of the piston,
the air is forced from the cylinder through the
communicating neck into the retort. For a
moment the mixture of oil-vapor and air is too
rich to explode, but as the piston progresses,
enough air is forced in to make the mixture
naturally explosive. This automatic explosion
is found to take place exactly as the piston is
making the next out stroke. Ignition takes

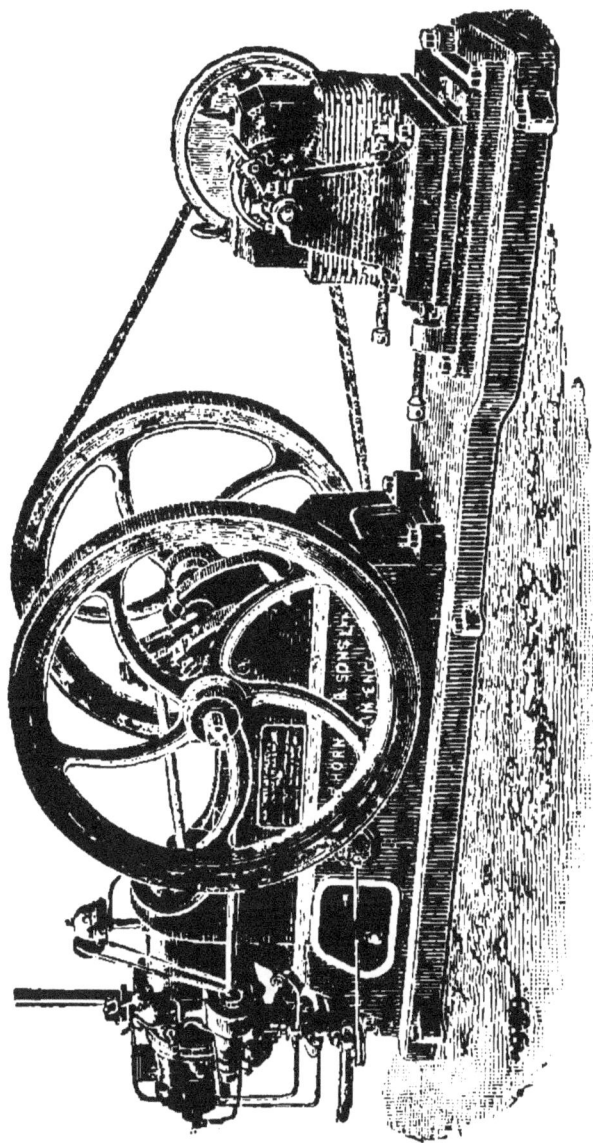

FIGURE 24.

place within the retort, the piston being protected by a layer of pure air. Analysis made of the gases indicate that the oil is completely burned in the cylinder with an excess of oxygen. The products of combustion formed consist mainly of steam and carbon dioxide diluted with nitrogen and oxygen, traces of carbon monoxide being also detected, so that the exhaust is not objectionable. The oil is sent to the retort by a small pump, which always supplies enough for the full power the engine is expected to produce. If less power is being used the speed increases a trifle, (less than 1 per cent.) and a high-speed "Porter" governor opens a by-pass and allows the surplus oil to return to a reservoir cast in the base of the engine. This engine gives first-class results and is largely used in England, but as far as the writer is aware is not yet on the American market, and knowing from actual experience in Birmingham, England, that its performance is very satisfactory, we will next describe an engine that will use any grade of oil or gasoline.

CHAPTER XIII.

The Birkholz or Raymond Improve Engine.

This engine is built by the T. I. Case Thresh-ing Machine Co. of Racine, Wis. The most important feature of this engine is the method of governing employed by means of which the number of explosions is always the same, but the strength of the impulse is varied by the strength of the gas and air mixture and the quantity that is admitted. In the factory these engines are tested to 25 per cent. over their normal power with a load of incandescent lamps. A single cylinder 2 h. p. engine of this make belted direct to a Warwick 2 K. W. dynamo held the needle of a Weston volt meter so steady that it was impossible to detect a quiver.

The engine is of the Otto or 4-cycle type, which, the makers claim is the simplest and best, as the piston does all the work.

It is built in single, double or quadruple cylinders, the multiple cylinders being employed
for economy of space and gas, neatness of de-
sign, and to lighten and divide the force of the
explosions. The makers state that they do not
depend upon multiplying the cylinders for
steady power, as a single cylinder engine will
regulate *almost* as closely as a quadruple one.
They consider that the vertical gas engine
has many points of superiority over the horizon-
tal type, the rocking motion being practically
done away with. The vertical type also econ-
omises space, a 50-H. P. not exceeding 5x7 ft.
of floor space. All sizes of engines are self-
starting. This engine is also one of the few on
the market provided with plugs to enable the
operator to take a card off his engine with an
indicator. The engine is shown in Figure 25.

FIGURE 25.

Illustrations.

Insert plate of working drawings and general dimensions of the Wadsworth-Warwick Gas Engine opposite page 40.

Index.

A

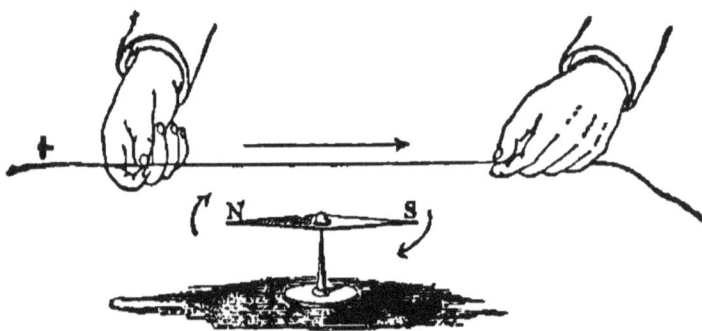

BUBIER'S
Popular Electrician.

A SCIENTIFIC ILLUSTRATED MONTHLY

For the Amateur and Public at Large.

Containing descriptions of all the new inventions as fast as they are patented; also lists of patents filed each month at the Patent Office in Washington, D. C. Interesting articles by popular writers upon scientific subjects written in a way that the merest beginner in science can understand. Also a Question and Answer Column free to all subscribers.

Price, Postpaid, $1.00 a Year.
Sample Copy, Ten Cents.

Send for it. You will be more than pleased.

Bubier Publishing Co., Lynn, Mass.

THE WONDERFUL
Pocket Vitagraph.

THE GREATEST NOVELTY OF THE AGE.

This little arrangement gives all the effects of the Kinetoscope. Objects move and people act as though they were alive. No. 1, Boxing Match. No. 2, Skirt Dance. No. 3, The Quarrel. No. 4, Street Fight. No. 5, Turkish Girl Dancer. No. 6, Serpentine Dance. No. 7, The Yankee Cop, showing two men fighting and the policeman comes along and arrests the wrong man. No. 8, The Funny Story, all the gestures, expressions, etc., plainly visible. Other subjects in preparation. Price, 10 cents each. Mail free. Anybody can work them, even a child.

To Agents, Salesmen & Dealers in Novelties.

AGENTS WANTED.—Liberal discounts will be made to them if they buy in quantities. Agents are now making $3.00 to $5.00 per day.

We can assure you it is the best selling article ever handled. A trial order will prove it. Sample 10 cents. We guarantee that you will be satisfied with it.

PRICES TO AGENTS AND DEALERS

50 C. PER DOZ. $5.00 PER GROSS. $25.00 PER THOUSAND.

Postage 10 cents per dozen extra. Expressage on gross lots or over to be paid by receiver. Terms: cash with order. Goods will be sent C. O. D. if one-fourth of the amount accompanies order. *Orders filled promptly.*

Bubier Publishing Co.,
LYNN, MASS.

EDITOR: BRIAN D. SZAFRANSKI, ELMA, NY USA -- JULY 15, 2016
COURTESY: WESTERN NEW YORK GAS & STEAM ENGINE ASSOCIATION
ALEXANDER, NEW YORK USA -- WWW.ALEXANDERSTEAMSHOW.COM

www.ingramcontent.com/pod-product-compliance
Lightning Source LLC
Chambersburg PA
CBHW021959190326
41519CB00010B/1329